It's Alive!

It's Alive! The Funniest Math Book Ever! combines the best of *It's Alive!* with the best of *It's Alive (and Kicking!),* plus 40 all new problems, to produce what is without a doubt the funniest math book ever! Teachers will especially appreciate the new edition's inclusion of a chart linking each problem with specific Common Core Standards, and another chart linking each standard to specific problems.

Marya took the kind of gooey, slimy, weird and disgusting science facts that kids love, and aided by her son and his best friend—both in middle school at the time—turned them into hilarious and engaging math problems. Simple in design but valuable in content, this book will thrill and excite your middle schooler by replacing typical, run-of-the-mill math problems with the stuff they're really thinking about, like:

☞ How many unknown-to-science microbes live in your belly button?
☞ What percent of refrigerators contain n.l.r.m.i.t. (no longer recognizable moldy icky things)?
☞ How many times is gas passed in the average 60-minute math class?
☞ How many gigabytes does your brain hold?

Math is fun when we get to figure the cost of a meal at the AfterMath Restaurant, with foods like Macaroni and Sneeze, Deep Fried Lint and Hot Sludge Sundae. Even the comprehensive answer key, including step-by-step detail and problem-solving strategies, is hilarious! Math and science combine with laughter in *It's Alive! The Funniest Math Book Ever!* Math class will never be the same!

Marya Washington Tyler, M.Ed., enjoyed teaching middle school students in Wisconsin, Idaho, Washington State and Alaska for 18 wonderful years. She continues to serve teachers by writing math books such as *Real Life Math Mysteries, It's Alive!, It's Alive (and Kicking!)* and *On the Job Math Mysteries,* all of which have been best sellers in the field of gifted education.

David Washington, now Partner Group Manager at Microsoft, with his best friend Asa, helped conceptualize the idea for the first edition and develop the problems while still in middle school.

Asa Kleiman, now a freelance illustrator, was the genius behind most of the humor you'll find in *It's Alive! The Funniest Math Book Ever!*

"Chock full of imaginative, fact-packed challenges that grab students' interest! *It's Alive! The Funniest Math Book Ever!* builds math and thinking skills while kids enjoy solving engaging puzzles."

Deb Fausti, *4th/5th grade Highly Capable teacher of 23 years*

"Quality teaching captures student interest, provides just the right challenge and adds spice with a bit of humor. Assignments connecting two or more curricula areas are golden. Quality teachers take advantage of classroom enthusiasm by encouraging student creativity to flow out of these lessons. Marya's book *It's Alive! The Funniest Math Book Ever!* offers plenty of material upon which to build quality teaching."

Hank Benjamin, *author of* Teaching in a Hyperactive Society *and a Dr. William Glasser National Quality Schoolteacher of the Year*

Including 77 WEIRD SCIENCE FACTS!

It's Alive!

The Funniest Math Book Ever!

Newly Revised and Updated Edition

Marya Washington Tyler, David Washington and Asa Kleiman

Illustrated by Eric Nelson and friends

Routledge
Taylor & Francis Group

NEW YORK AND LONDON

Illustrated by Eric Nelson*
*The cover and problems 1, 2, 3, 4, 6, 8, 9, 11, 19, 20, 23, 24, 31, 36, 40, 43, 47, 50, 53, 57, 64, 68, 72, 83, 86, 97, 105, 108, and 117 were drawn by David's family

Second edition published 2025
by Routledge
605 Third Avenue, New York, NY 10158

and by Routledge
4 Park Square, Milton Park, Abingdon, Oxon, OX14 4RN

Routledge is an imprint of the Taylor & Francis Group, an informa business

© 2025 Marya Tyler, David Washington and Asa Kleiman

The right of Marya Tyler, David Washington and Asa Kleiman to be identified as authors of this work has been asserted in accordance with sections 77 and 78 of the Copyright, Designs and Patents Act 1988.

First edition published by Prufrock Press 1996

ISBN: 978-1-032-51014-9 (pbk)
ISBN: 978-1-003-40590-0 (ebk)

DOI: 10.4324/9781003405900

Typeset in Times New Roman
by Deanta Global Publishing Services, Chennai, India

To the One who does all things well...   Thank You!!

About the authors

Asa and David are two computer geeks, sufficiently odd to be considered peculiar by everyone except David's dog.

Asa Kleiman was born from alien species, but hides it relatively well.

David Washington is normal in comparison. David's talents include swallowing a can of soda while hanging upside down, but he denies the frog incident entirely.

Marya has taught in middle school for lots of years, and she's David's mom.

Eric Nelson is an erstwhile illustrator and tintypist from Chili, Wisconsin, whose funny bone loved drawing the illustrations for these goofy situations.

Standards for each problem

(see also **Problems for each standard** in the next section)

1.	**Please don't**	using area to find length and width, multiplication
2.	**Slosh**	fraction—multiplying and simplifying, converting cups/gallons
3.	**What's all that racket!**	converting feet/miles, division multi-digit whole numbers
4.	**Ice pops**	adding negative and positive numbers, temperature
5.	**I'll be with you in a jiffy**	ratio, converting units of time
6.	**Doggie menu**	2-step problem, create equation
7.	**It wasn't me!**	2-step problem, finding mean
8.	**Crayfish smoothie**	2-step problem, dividing whole numbers, subtraction of decimals
9.	**Popcorn gone bad**	converting pounds/tons, dividing multi-digit whole numbers
10.	**Crumbly**	2-step word problem
11.	**Shrimp, huh?**	division whole number, creating a bar graph
12.	**PFFFFFFT**	multi-step problem, finding mean, division by decimal, metric
13.	**Dandelion doughnuts**	converting decimal to percent, rounding to hundredth, extension
14.	**I'm coming as a dung beetle**	finding percent of, multiplying by decimal
15.	**Fly tracks**	finding perimeter
16.	**Close the door!**	multiplication, 2-step word problem
17.	**Stop staring!**	finding area/square inches, multiplying by decimal, 2-steps

18.	**Bring in the SWAT team**	multiplication large numbers, 2-step word problem
19.	**Care to go for a swim?**	converting cups/quarts, 2-step problem
20.	**Star pupil**	2-step problem, converting seconds/minutes
21.	**Cheese!**	addition of decimals
22.	**Lawn ornaments**	adding decimals, percents, interpreting & analyzing results
23.	**It's normal**	rate, 2-step problem, converting days/hours
24.	**But it was good dirt**	3-step problem, multiplying decimals, converting days/years
25.	**Stop the car!**	2-steps, rate, multiplying decimals, converting hours/minutes
26.	**Let me in!**	rate, multiplication
27.	**Drowning in droplets**	multiplication
28.	**Eliminate waste**	2-step problem, rate, fraction, converting days/hours
29.	**Pooky**	finding percent of, multiplying decimals, converting hours/minutes
30.	**Bats in the tent**	division large numbers
31.	**Flapflapflapflap**	ratio, rate, converting minutes/seconds
32.	**Thank you, garbage people**	multiplication by decimal, extension
33.	**Did you remember to flush?**	division large numbers
34.	**Slimy**	2-step problem, rate
35.	**Tern, tern, tern**	multiplication large numbers
36.	**Zit city**	multiplying large number by decimal
37.	**Look closely**	multi-step, converting teaspoons/tablespoons/cups/gallons

38.	**Worm digging**	multiplication large number by fraction or decimal
39.	**Panda poop**	multiplication by fraction, simplifying fraction
40.	**Ants rule**	converting very large numbers
41.	**Hot dog**	percent to decimal, multiplying by decimal, metric
42.	**Snarly**	2-step problem, subtraction/ multiplication decimals
43.	**Supersweeper**	ratio, large numbers, simplifying fraction
44.	**I don't get it**	multiplying by decimal
45.	**Where did my mailbox go?**	2-step problem, rate, fraction, converting hours/minutes
46.	**Skin scales 25 cents a pound (Do I hear 30!?)**	converting ounces/pounds, multiplying by decimal
47.	**Smile!**	rate, feet/yards/miles, multiplying/ dividing decimals
48.	**I thought I smelled something**	division by decimal
49.	**Coffee, anyone?**	multi-step problem, converting ounces/cups/pint
50.	**A very hairy ordeal**	rate, multiplying decimals, rounding decimals, metric
51.	**Locker room**	3-step problem, finding mean, large numbers
52.	**Nematode sea worms**	converting mm/meters/km, equation, very large numbers, metric
53.	**Adventure awaits**	multiplication, creative problem solving
54.	**Classical Kazoo Symphony**	multiplication
55.	**Mirror, mirror, on the wall**	multi-step, percent of, multiplying by decimal, day/hour/minute
56.	**After School Science**	multi-step problem, make a chart, equation, converting hr/min/sec

57.	**Human vs. pineapple**	2-step finding percent of, changing decimal to percent
58.	**You're next, young man**	3-step problem, cm/mm, multiplying decimals, metric
59.	**How many didn't bite?**	2-step problem, very large numbers
60.	**Great-grandma's knitting**	2-step problem, multiplying fractions, subtracting fraction
61.	**Whole-wheat rutabaga caramel tofu surprise**	multi-step complex problem, writing equation
62.	**Fungus among us**	multiplying fractions/decimals, finding area, interpreting answer
63.	**Fuzzy apple logic**	finding percent, multiplying decimal, converting decimal/percent
64.	**Let us see your hair, David**	make a list, find a pattern, add/subtract fraction
65.	**Shhhhh**	3-step problem, multiplying by decimal, multiplying by fraction
66.	**N.L.R.M.I.T.**	3-step problem, convert fraction/decimal, multiplying by decimal
67.	**Sludge**	3-step problem, volume—finding depth, writing equation,
68.	**No, we are not buying a 3D TV**	2-step problem, understanding decimals, multiplying by decimal
69.	**Camp Torture**	2-step problem, adding fractions, common denominator
70.	**Heinz and Gunter**	subtracting fractions, common denominator
71.	**Gum on the seats**	2-step problem, multiplying large number by fraction
72.	**Chicken noodle sloop**	multiplying by fraction
73.	**Candy jar**	multiplying by fraction
74.	**People shed too**	rate, days/hrs/min, multiplying decimals, very large numbers

75.	Lumps	2-step problem, converting fraction to decimal to percent
76.	Taj Mahal	finding volume (cubic feet), multiplication
77.	Longest toe	gathering data, creating bar graph
78.	An acre of pizza, please	multi-step problem, seconds/minutes/hours; rate
79.	Pair of socks	converting milligrams/grams, days/years, metric
80.	School janitor	finding ratio, reducing fraction, finding percent
81.	Overflow	3-step problem, divide by decimal
82.	Cans of Spam™	2-step problem, converting pounds/tons, dividing by fraction
83.	Old, fuzzy grapes	multiplying fractions, finding percent of, interpreting decimal
84.	Frog eggs	large numbers, analyzing answer
85.	Eyelash mites	multi-step problem, rate, converting cm/mm, metric
86.	Cow gas	2-step problem, converting pints/quarts/gallons
87.	Lucky Charms™	2-step problem, converting months/days
88.	Cereal boxes	complex problem, converting year/months
89.	Back yard fertilizer	complex problem, area, multiplying by fraction, fraction to percent
90.	Taste buds	3-step problem, finding percent, multiplying by decimal
91.	Grocery store bulletin board	creating original word problems
92.	Leeches	3-step problem, converting pounds/ounces
93.	Royal Canadian Bagpipers	2-step problem, adding/subtracting time, fractions

94.	**Do you smell something?**	finding area of a circle, using area formula, multiplying decimal
95.	**Tick eating habits**	2-step problem, ratio, converting days/years, dividing by fraction
96.	**What would you do with $3,500?**	creating original word problems
97.	**ASA**	3-step problem, subtraction/ multiplication/division of decimals
98.	**Glub**	rate, converting day/yr, dividing by decimal, very large numbers
99.	**Mucus moves**	2-step problem, rate, converting m/ mm, hr/min, metric
100.	**Cellmates**	complex problem, make a chart
101.	**Asa's airplanes**	3-step, rate, convert days/hrs, mm/ cm, multiplying fraction
102.	**Up, up, and away**	3-step, rate, pints/quarts/gallons, dividing by decimal, rounding
103.	**Floss 'em**	5-step problem, converting inches/ feet, dividing fractions
104.	**Saliva mouth**	5-step problem, convert pt/qt/gal, days/hrs/min, divide fraction
105.	**What's that in my water bottle?**	2-step, radius/diameter/ circumference/pi, multiply decimal
106.	**Breathe in**	3-step problem, volume, ft/cu yd, multiply by fraction
107.	**Arteries**	draw a diagram, look for a pattern, divide large numbers
108.	**Live with Asa!**	probability, combinations
109.	**A rat can fall**	draw and analyze diagram, multiply by decimal
110.	**Censorship!**	complex problem, write an equation, multiplying by percent,
111.	**Smork**	2-step problem, convert hrs/min, simplify fractions, find ratio

Problems for each standard

Adding fractions	69
Adding negative and positive numbers	4
Analyzing/interpreting answer	83, 84, 112
Area of circle—finding	94
Area of rectangle—finding	17, 62, 89
Area—using to find length and width	1
Averages (mean)	7, 12, 51
Bar graph creating	11, 77
Chart/map/diagram—make a	56, 64, 100, 107, 109
Circle—area finding	94
Circle—circumference, radius, diameter, Pi	105
Conversions milligrams/grams	79
Conversions ounces/pounds/tons	9, 46, 82, 92
Converting cups/ounces/pints/quarts/gallons	2, 19, 37, 49, 86, 102, 104
Converting decimal to percent	118
Converting decimal/percent	41 57, 63, 65
Converting fraction to decimal	66, 115
Converting inches/feet/yards/miles	3, 47, 53, 103, 114
Converting metric/English	113
Converting mm/cm/meters/km	52, 58, 85, 99, 101
Converting time—sec/min/hr/day/mo/yr	5, 20, 23, 24, 28, 29, 31, 34, 45, 55, 56, 79, 88, 98, 114
Converting—large numbers	40
Creative problem solving	53, 91, 96
Decimal/percent	41 57, 63, 65, 75
Decimals—addition/subtraction	8, 21, 22, 42, 97, 119

Time—converting minutes/hours	29, 45, 55, 56, 74, 78, 99, 104, 111, 114
Time—converting seconds/minutes	5, 20, 31, 56, 78, 114
Two-step/multi-step problems	6, 8, 10, 12, 51, 59, 61, 68, 71, 74, 78, 81, 82, 84, 86, 87, 88, 89, 90, 92, 93, 95, 99, 100, 101, 103, 104, 106, 110, 112, 113, 114, 117, 119
Volume cups/ounces/pints/quarts/gallons	2, 19, 37, 49, 86, 102, 104
Volume teaspoons, tablespoons, cups	37
Volume—cubic feet, cubic yard	68, 76, 106
Volume—sphere	115
Weight converting milligrams/grams	79
Weight converting ounces/pounds/tons	9, 46, 82, 92

What's with all the big numbers?

How do you say a number like this?
999,888,777,666,555,444,333,222,111,000

# of commas	You Say	# of commas	You Say	# of commas	You Say
1	thousand	9	octillion	16	quindecillion
2	million	10	nonillion	17	sexdecillion
3	billion	11	decillion	18	septen-decillion
4	trillion	12	undecillion	19	octodecillion
5	quadrillion	13	duodecillion	20	novem-decillion
6	quintillion	14	tredecillion	21	vigintillion
7	sextillion	15	quattuor-decillion		*and on and on...*
8	septillion	10	nonillion	33	Googol

So you say, "999 octillion, 888 septillion, 777 sextillion, 666 quintillion,
555 quadrillion, 444 trillion, 333 billion, 222 million, 111 thousand".

IMAGINE:

A HUNDRED peanut butter cups (100)
Now imagine TEN HUNDRED peanut butter cups. That's a thousand (1,000).
Imagine A THOUSAND THOUSANDS! That's a million (1,000,000)!
And THOUSANDS of THOUSAND THOUSANDS!
That's a billion (1,000,000,000)!

DOI: 10.4324/9781003405900-1

BIG NUMBERS are handy when you're talking about stars, cells and computers.

KILO means thousand.
MEGA means million.
GIGA means trillion.
TERA means quadrillion.

PETA means quintillion.
EXA means sextillion.
ZETTA means septillion.
YOTTA means octillion.

Note to teachers

If someone in your class asks, "May we use calculators?"
Our answer is, "Are mice rodents? Is Asa's sister annoying?
Does 1 + 1 = 2?"
Yes, you can use calculators.

Problems

The bold facts in this book are true.

(Everything else is horsefeathers.)

1. Please don't

FACT: The largest human organ is the skin, with a surface area of about 25 square feet. If all the skin on your body was laid out square, what would its length and width be?

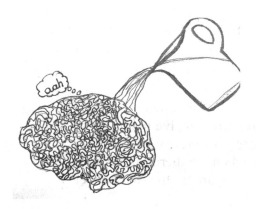

2. Slosh

FACT: The human brain contains about 2 pounds of water. A gallon of water weighs about 8 pounds. There are 16 cups in a gallon. How many cups of water are in the brain?

DOI: 10.4324/9781003405900-2

3. What's all that racket!

FACT: The Eurasian crane flies higher than any other bird, reaching 32,000 feet! At that altitude, they are invisible from the ground, but their calls are so loud, they can be heard from the ground (which is why it's a good thing they fly so high!) Approximately how many miles high do they fly? *(5,280 feet = 1 mile)*

4. Ice pops

FACT: The Siberian salamander can survive in temperatures as low as −58 degrees F. Meanwhile, David's sister Carrie turns into a popsicle when the thermometer is set at 68 degrees F. How many degrees apart are the salamander popsicle and the sister popsicle?

YOUR ARTIST WILL
BE BACK IN
A JIFFY

— Nelson '23

5. I'll be with you in a jiffy

FACT: A jiffy is an actual measure of time. There are one hundred jiffies in a second. There are one thousand milliseconds in a second. How many jiffies occurred in the 12 seconds it took you to read this? How many milliseconds?

6. Doggie menu

The vet advised David to feed Scruffy 1,500 calories per day. Already today, Scruffy has eaten one bowl of dog food (325 calories), one doggie treat (20 calories), ½ cup Cheepios which David's sister spilled on the floor (50 calories), half a peanut butter sandwich out of David's lunch bag (170 calories), one yucky thing Scruffy found on the ground (40 calories) and one butterfly (1 calorie). How many calories will Scruffy still need to consume today? *(Create an equation and solve.)*

7. It wasn't me!

FACT: The average person passes gas 14 to 23 times a day.
If David passes 14 times and the kid who sits in front of David passes 23 times, what is the mean?

8. Crayfish smoothie

FACT: Boys require an average of 2,800 calories per day. Girls require an average of 2,200 calories per day. One crayfish provides 70 calories. Because crayfish are invasive in Wisconsin, Asa convinces his class to join him eating nothing but crayfish on Friday. When Nothing But Crayfish Day rolls around, how many more crayfish should Asa eat than his identical twin sister (to the nearest tenth of a crayfish)?

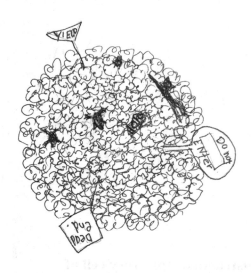

9. Popcorn gone bad

FACT: The employees of The Popcorn Factory made a popcorn ball that weighed 3,423 pounds. If it started rolling and didn't stop rolling until it had flattened four street signs, three toads and one innocent gray squirrel (who somehow all survived without injury), how many tons did this thing weigh anyway? *(1 ton = 2,000 pounds)*

10. Crumbly

David's grampa spills 17 crumbs on the floor each meal. If he eats 3 meals today, and Scruffy laps up 42 of his crumbs, how many crumbs are still on the floor at the end of the day when David's mom mops?

11. Shrimp, huh?

FACT: Chromosomes are the instructions for every cell of your body. Every species has a set number of chromosomes. Peas get along fine with 14 chromosomes. Humans 46. There are 80 chromosomes in pigeons. Humans have about half the chromosomes of shrimp. Make a bar graph showing the comparison.

12. PFFFFFFT

FACT: Each day, the average person produces about 500 to 1,500 milliliters of gas, and expels it in 16.5 farts, most of which are odorless. Using the average amount of gas produced per person per day, find the average amount of gas expelled per poot. *(Round to the nearest tenth.)* (By the way, how does one fart half a fart?)

Doughnut HOLE

13. Dandelion doughnuts

FACT: At least 80,000 plant species are edible, but most people only eat 30 of them. What percent of edible plant species aren't being eaten? (Stinging nettle soup, anyone?) *(Round to nearest hundredth of a percent.)*

EXTENSION: Make a list: How many different plant species do you eat?

14. I'm coming as a dung beetle

FACT: About 80 percent of the world's living species are insects. If everyone at your school is dressing up as a different animal for the upcoming Halloween party, and there are 600 people in the school, how many should come dressed as insects?

15. Fly tracks

How far did a fly, which nibbled its way around the perimeter of David's mom's 9″ x 13″ prune-frosted tofu grapefruit cake, walk?

16. Close the door!

Asa looks in the refrigerator 18 times a day and spends an average of 8 seconds looking each time. David looks in the refrigerator 12 times a day and spends an average of 12 seconds looking each time. Who spends more time looking in the refrigerator, and how much more time does he spend per week?

17. Stop staring!

FACT: As many as 5 million mites have been found crowded together in one square inch. How many mites could be sitting on this 8½" x 11" page watching you right this minute?

18. Bring in the SWAT team

FACT: During one summer alone a single pair of houseflies, under ideal conditions, can produce as many as 335,923,200,000,000 offspring. A single pair of houseflies is buzzing around reproducing under ideal conditions. If it takes Asa 12 swats to kill one fly, how many swats will he need to make to get rid of all the flies produced from that single pair in one summer?

19. Care to go for a swim?

FACT: The average human produces 25,000 quarts of spit in a lifetime, enough to fill two swimming pools. How many cups of spit would fill one swimming pool? *(4 cups = 1 quart)*

20. Star pupil

Scruffy was a big hit at doggie obedience class…at least among other dogs. Ten seconds from the start he began howling: AWOOOO! Ten seconds later another dog joined in. AWOOOO! Every ten seconds another dog joined in until all 21 dogs were howling. AWOOOO! How many minutes and seconds had passed when all 21 dogs were howling? AWOOOO! (And is it fair that Scruffy was not invited back?)

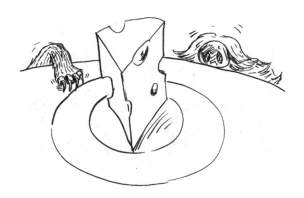

21. Cheese!

As soon as David's dog Scruffy hears the crinkle of cheese being opened, he leaps from a sound sleep and is at David's feet in 0.84 seconds hoping for a bite of cheese. The sound of the peanut butter jar opening takes Scruffy a quarter second longer. How long does Scruffy take to come for peanut butter?

22. Lawn ornaments

David and Asa have calculated the percent of houses in their town having each type of lawn ornament.

44.0 percent—pink flamingos
24.0 percent—goose in Green Bay Packer uniform
19.2 percent—gnomes
17.2 percent—cutesy squirrels and bunnies
14.5 percent—life-size deer (dressed in orange during hunting season)
0.9 percent—crystal balls
0.4 percent—little boys fishing

Determine the sum of these percents. How is that possible?

23. It's normal

FACT: Scientists say it is considered quite normal to pass gas 15 times in a 24-hour period (quietly, please). In a 60-minute math class consisting of 27 people and one teacher, how many times would this normally occur?

24. But it was good dirt

FACT: The average person ingests about 100 milligrams of dirt a day. (One milligram is $\frac{1}{1000}$ of a gram. One gram weighs about the same as a raisin.) How many grams of dirt have you ingested in the last ten years? *(Use 365.25 for the number of days in a year.)*

25. Stop the car!

FACT: A full bladder fills at a rate of about 2 milliliters per minute. At 400 ml of liquid, it's full. David and his mom just left a rest stop where David used the restroom and then drank almost a liter of water. If it's three and a half hours to the next rest stop, will David make it to the restroom before his bladder is completely full and he explodes? *(Show your work.)*

26. Let me in!

As it turned out, David's mom stopped the car by the side of the road, and David quickly ran into the woods. The entire operation only took 54 seconds, during which time he was bitten by 17 mosquitoes per second. David returned to the car with how many mosquito bites?

27. Drowning in droplets

FACT: The average person expels 300 saliva droplets per minute while talking. How many saliva droplets did your teacher expel during the last 55-minute lecture?

28. Eliminate waste

FACT: It takes a full day and a half for waste to travel the 36 inches through the colon before it is eliminated. How many inches per hour is that? Use fractions if necessary.

29. Pooky

FACT: The average cat spends 85% of her day apparently doing nothing. If David's cat, Pooky, is relatively average in this regard, how many hours and minutes does Pooky have to do other cat things like sharpening claws on furniture, knocking dishes off the counter and digging up potted plants?

30. Bats in the tent

FACT: One little brown bat can catch and eat 500 mosquitoes in a single night. While Asa and David were camping at Swampy Hollow State Park, 135,000 mosquitoes entered their tent. How many little brown bats should they invite into their tent?

31. Flapflap flapflap

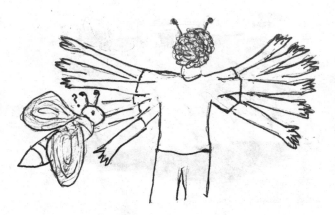

FACT: Honeybees flap their wings 230 times per second. It took Asa 2 minutes and 20 seconds to flap his arms 230 times, at which point he fell off the porch into the thornbush. How many times faster is the bee?

32. Thank you, garbage people

FACT: The U.S. Environmental Protection Agency (EPA) estimates that the average American throws away 4.4 pounds of trash per day. For a school project, David and Asa tried carrying around all their trash, but the teacher made them quit after three days because of the stink. How much did each boy's trash weigh?

EXTENSION: Find something else that weighs about the same.

33. Did you remember to flush?

FACT: NASA spent $23.4 million on a new and improved toilet for astronauts on the International Space Station.
Assuming the toilet has been used 121,875 times—we were unable to locate any actual statistics on this—and assuming the astronauts always remembered to flush, how much has the Space Station toilet cost per flush?

34. Slimy

Asa just finished his 7th week working 6 days a week as a lifeguard at a pool. His job includes cleaning the pool filter, and so far he has found 21 slimy hair scrunchies in a variety of colors. If he continues finding hair scrunchies at this rate for the remainder of the summer (one more week), what is the total number of slimy hair scrunchies he will have found?

35. Tern, tern, tern

FACT: Arctic terns fly the round trip from the Arctic to the Antarctic each year. That's 11,000 miles one way. If Arctic terns could collect one frequent flier mile for every mile they fly, how many frequent flier miles could an Arctic tern collect over a 20-year period?

36. Zit city

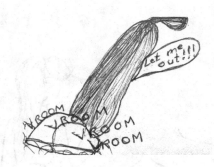

FACT: There are 2½ times as many oil glands on your skin as there are hairs on your body. If there are 200 million hairs on your body, how many zits will you have if every oil gland suddenly clogs with oil and dirt and breaks forth into a zit?

37. Look closely

FACT: Dust mites eat dust and dandruff from our bodies, which is valuable to our bodies and the ecosystem. One half teaspoon of dust contains about a thousand mites. A two-gallon vacuum bag filled with dust contains how many of these helpful little mites? *(1 gallon = 16 cups) (1 cup = 16 tablespoons) (1 tablespoon = 3 teaspoons)*

38. Worm digging

FACT: One acre of good soil usually contains about 3 million worms. Asa and David are digging for worms on a neighbor's ¾-acre neatly mowed, meticulously maintained front lawn. How many worms can Asa and David expect to find? (Assuming the owner doesn't see them first.)

39. Panda poop

FACT: Pandas basically only eat bamboo, and they eat about 30 pounds of the stuff each day. Bamboo is very high in fiber, so pandas poop out $\frac{4}{5}$ of what they take in each day. How many pounds of poop does a panda poop per day?

40. Ants rule

FACT: 20 quadrillion. That's how many ants there are on the Earth at any given time. 20 quadrillion ants is the same as 20,000 trillion ants, or how many billion ants? (And what is that crawling up your neck?)

41. Hot dog

If school lunch hotdogs bounce 12 percent of the height from which they fall, how high would one bounce if Asa and David dropped one from:

a. the top of Asa's garage—12′?
b. the top of City Hall—85′?
c. a Boeing 747—12,000 meters

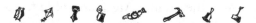

42. Snarly

Beginning on November 1st, David's dog Scruffy accumulates 15 new tangles in his fur per day. However, in one 24-hour period he will only hold still long enough for David to remove 14.75 of his tangles. At this rate, how many tangles will Scruffy have on May 1st (181 days later) when it is finally warm enough to give him a doggie haircut? *(Round to the nearest hundredth.)*

43. Supersweeper

FACT: **It's estimated that there are more than 100 million pieces of space junk traveling uncontrolled through orbit that could cause collisions in space and even render future space exploration impossible.** David and Asa have invented a supersweeper which they name Aardvark, able to suck up 100 pieces of space junk while positioned in their back yard. What fraction of the space junk has it collected?

44. I don't get it

FACT: **Humans spray an average of 2.5 saliva droplets into the air with every word they speak.** How many saliva droplets are released into the air every time someone in your math class says, "I don't get it"?

45. Where did my mailbox go?

FACT: David's sister is getting her driver's license. If she hits a mailbox every 15 minutes, what will her grand total be after 2½ hours of driving practice?

46. Skin scales 25 cents a pound (Do I hear 30!?)

FACT: A heavy skin shedder sheds 1½ grams of skin a day, or about 1 pound a year. If skin scales were selling for 25 cents an ounce, how much could someone like David, who is a pretty heavy shedder, make by selling his skin scales every year? (If you find a buyer, please let us know.) *(16 ounces = 1 pound)*

47. Smile!

FACT: A rattlesnake travels at his top speed of two miles per hour. Right now, one is slithering the 880 yards over to the warm rock where you're sunbathing, so that it can sunbathe next to you. How long until you look over and see his smiling face? *(1 mile = 5,280 feet) (3 feet = 1 yard)*

48. I thought I smelled something

One time David brought a little baggie of sardines for lunch, which fell to the bottom of his locker uneaten. David then opened his locker a total of 147 times before he finally noticed the sardines while searching for his gym socks. If, on average, David opens his locker 3½ times a school day, how many school days later was it when he finally found (and thankfully discarded) the fish?

49. Coffee, anyone?

FACT: Each person sweats about 1½ pints a day. How many 8-ounce coffee mugs would this fill? *(8 ounces = 1 cup) (2 cups = 1 pint)*

50. A very hairy ordeal

If you shaved your head and did not cut it again, and it grew at the rate of 0.35 mm per day, how long would your hair be in 15 years? Give answer in meters rounded to one decimal place. Draw a picture showing approximate length to scale. *(365.25 days in a year)*

51. Locker room

FACT: Humans have 2,000,000 to 5,000,000 sweat glands. After your next gym class of 35 people, what will be the average number of sweat glands going into the locker rooms?

52. Nematode sea worms

FACT: There are approximately 40 septillion nematode sea worms in the world, and although slithery, they are actually quite cute. If all nematode sea worms (average length 2 mm) joined end to end in the Annual Nematode Seaworm Synchronized Swim Show, how many kilometers long would the line be?

53. Adventure awaits

FACT: If you took all the highways in the US and laid them end to end around the equator, some people would get really mad. A lot of people use those roads to get to work and stuff. And you use them to get to school. Considering there are 180 school days in the average school year, how many miles have you traveled back and forth between school and home? If you took all those trips and laid them out in a line, what's the nearest cool place you could get to? *(Google Maps will help.)*

54. Classical Kazoo Symphony

While unsuspectingly watching TV, Asa finds out that the Live-and-In-Concert Classical Kazoo Symphony Orchestra and Marching Band has come on. It takes him 4 minutes to dig through each box of junk in his room to find his remote. There are 14 boxes of junk in Asa's room. He finally finds his remote in the last box. How much of the Classical Kazoo Symphony Orchestra and Marching Band was he subjected to?

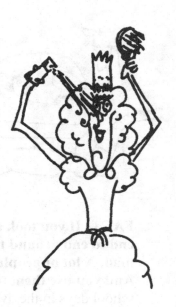

55. Mirror, mirror, on the wall

David figures that his older sister spends 14 percent of her waking life every day in front of the mirror. If she sleeps 9 hours a day, how many hours and minutes does she spend in front of the mirror?

56. After School Science

Asa and David's After School Science club will be showing the following videos on Saturday:

a. "Your Spleen and Why You Should Care" (5 hours 45 seconds)
b. "Science Wonders: The Varied Sounds of Passing Gas" (3 hours 17½ minutes)
c. "Studies in Nasal Drip" (86 minutes)
d. How many hours, minutes and seconds will you have to sit through to see all three?

57. Human vs. pineapple

FACT: Chromosomes contain all the instructions for the growth and functions of your body. Every species has a set number of pairs of chromosomes. Humans have 23. Pineapples have 25. The pineapple instruction manual is what percent larger than the human?

58. You're next, young man

While waiting for the dentist, David bit off 5 percent of each fingernail. His fingernails each used to be 1.2 cm long. How many total millimeters of fingernails did he bite off?

59. How many didn't bite?

FACT: It is estimated that there are 25 million assorted insects hanging in the air over every square mile of the continental United States. During his first week attending Camp Torture (10 square miles of Wisconsin swampland), Asa was bitten by 1,738 insects. Asa's counselor comforted him by saying, "Look, kid, just think how many insects didn't bite you." How many Camp Torture insects did not bite Asa?

60. Great-grandma's knitting

David's great-grandma was $\frac{3}{4}$ done with her knitting project when David's shoe caught on the yarn. By the time she got his attention, ½ of what she had done had unraveled. Now how much more does she have to knit in order to finish the project?

61. Whole-wheat rutabaga caramel tofu surprise

Asa and David were desperately hungry and between them accidentally ate 11 slices of Ms. Washington's whole-wheat tofu rutabaga caramel surprise (with 13 tofu chunks per slice). If Asa ate 39 more chunks than David, how many slices of whole-wheat tofu rutabaga caramel surprise did they each eat?

62. Fungus among us

FACT: A fungus found in Washington state is considered the largest living thing on Earth. It covers 2.5 square miles. Asa and David are trying to get a cutting of the fungus. If they can get it to grow that big in their local park, which measures 1½ mile by 1½ mile, how much of the park will be covered?

63. Fuzzy apple logic

If David's sister leaves an apple core under her bed and it decomposes into a mass of fuzzy gray mold, losing 4 percent of its original mass each day, what percent of the original apple will be left when David's mom finds it two weeks later?

64. Let us see your hair, David

David's mom has taken to cutting his hair herself. David wore his Oyster Shuckers Bowling League hat for three straight days after his last haircut. The one before, he wore his Mime Radio Fan Club cap for 4½ days straight, and after the haircut before that one, he wore his Camp Torture hat for six straight days.

According to the pattern, is his mom getting better or worse, and how many days will David have to wear his Marshfield Lawn Ornament Society (MLOS) hat after the next haircut?

65. Shhhhh

Asa figures he has about 320 ideas a day, but at least 20 percent of these ideas get interrupted—either by his sister, his other sister, his other sister, his mom, the dog, the dog's friends or the phone. Of the interrupted ideas, ¼ return later. According to these estimates, how many ideas does Asa lose every day?

66. N.L.R.M.I.T.

If 70 percent of refrigerators in the United States contain n.l.r.m.i.t. (no longer recognizable moldy icky things), and 7 out of 8 of these contain more than one n.l.r.m.i.t., what percent of refrigerators contain more than one n.l.r.m.i.t.?

67. Sludge

FACT: The 25′ × 50′ septic pool at the wastewater treatment plant contains 157,000 cubic feet of sludge. *(Volume = length × width × depth.)* **Once a wastewater sanitation worker tripped and fell into the septic pool.** Fortunately, he was able to climb right back out without sinking far, but how deep is the sludge pool?

68. No, we are not buying a 3D TV

David's mom is shopping for furniture. Asa and David are helping her (mistake number one). How much would each of these items cost, after adding the $40 delivery fee and then adding the 5 percent Wisconsin sales tax? (Round to the nearest cent.)

☞ $1,039.99 Genuine 100 Percent Cathair Couch—specify tabby or Siamese. (No animals were harmed in the making of this couch.)

☞ $1,995.99 Overstuffed Quicksand Couch—soft, very soft.

☞ $1,123.58 Giant Whoopee Cushion Bean Bag Chair.

☞ $1,421.06 Prestained Wonder Couch—your choice of catsup, mustard, grease, chocolate, vomit or ink.

☞ $1,475,380.64 Lazy Person Couch—comes with refrigerator/ freezer, microwave, bidet, back-up generator, robot servant, central air and heating.

☞ Act now and receive free back scratcher and bottomless chips.

For each price add the $40 delivery fee. Then you could multiply by 0.05 and add the price, but it's easier to just multiply the price by 1.05. Then round to the nearest cent.

69. Camp Torture

At Camp Torture, where David and Asa each spent a few weeks last summer, there was a rule that if you don't finish your lunch, you never eat again. Ever. Under the bushes out behind the cafeteria/infirmary on Friday evening, Asa and David discovered the following sandwich remains:

a. ⅜ egged salad sandwich

b. ¼ artificial preservative spread on no-wheat whole bleach bread

c. ⅜ llama milk cheese on whole brain sneet bread

d. ⅝ knuckle sandwich

e. ⅞ toe-food and toe-mato on hole wheat crisps

f. ½ Salmon Ella sandwich

g. ¾ lox and keys on unleveled bread

h. ¼ Soilent Green Burger Deluxe

i. When they pieced together all of the sandwich parts, how many total sandwiches had they found?

70. Heinz and Gunter

David and Asa then threw the sandwiches to Heinz and Gunter, the camp Dobermans, who were distracted by eating the sandwiches (and later by indigestion) long enough to allow David and Asa to slip by unnoticed. When the boys returned, they found that together the dogs had eaten exactly ¼ of a whole sandwich from each type of sandwich remaining. What part of each type of sandwich was now left?

71. Gum on the seats

FACT: Yankee Stadium seats 52,326 spectators. At the last baseball game, David and Asa's independent research found gum stuck to the bottom of 2 out of 3 seats, in a variety of flavors and colors. If their research is correct, how many seats in Yankee Stadium have gum stuck to them, and what flavors are they?

72. Chicken noodle sloop

FACT: Two-thirds of skin scales examined in the laboratory are found to be inhabited with large colonies of bacteria.
If your school lunch cook absentmindedly scratches 750 skin scales off into the lukewarm chicken noodle soup, how many skin scales containing large colonies of bacteria just dropped into the soup? (Of course, this does not happen in your school.) (Hopefully.)

73. Candy jar

At Asa's house there is a candy jar in the living room which is ⅓ full of good fruit candies and ⅔ full of revolting butterscotch candies. How many revolting butterscotch candies should Asa expect to find in any handful of 12 candies?

74. People shed too

FACT: About 10 billion skin scales peel off each of our bodies every day. How many skin scales can you expect to peel off your body in the next 20 minutes? (Kindly refrain from shedding on the book.)

75. Lumps

Asa and David poured a 48-ounce bowl of David's grama's gravy through a strainer to remove the lumps. After straining, the gravy weighed in at 26 ounces. What percent of the gravy would not go through the strainer? *(Round to nearest percent.)*

76. Taj Mahal

If David's front yard measures 110′ x 50′ and is tightly packed with snow 2 feet deep, how many cubic-foot bricks of snow could he use to build a scale model of the Taj Mahal?

77. Longest toe

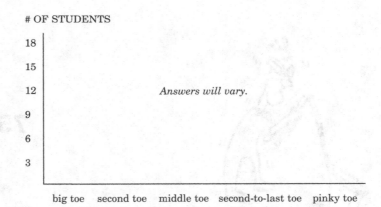

FACT: David's big toe is longer than his other toes. Asa's middle toe is the longest. Which one of your toes is the longest? List the students in your class, and record next to each name which of their toes is the longest: BIG TOE, 2nd TOE, 3rd TOE, 4th TOE or PINKY TOE. Then use graph paper to create a bar graph showing the number of people for each toe.

78. An acre of pizza, please

FACT: Someone who didn't have anything better to do determined that US citizens eat approximately 100 acres of pizza per day, which is about 350 slices per second. Assuming there are eight slices in a pizza, how many whole pizzas are eaten per half hour?

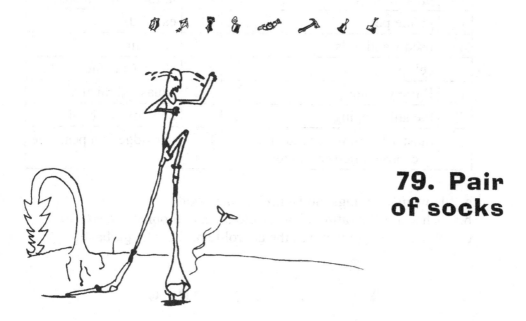

79. Pair of socks

FACT: The average pair of socks traps about 190 milligrams of skin a day. If Asa wears his socks two (non-leap) years straight without changing them, how many grams heavier will they be when he finally takes them off?

80. School janitor

The school janitor has been keeping a record of things he finds on the floor when he sweeps up after school. On Tuesday of this week, he found:

84	hair ties	42	fuzzy wads of gum
49	chewed pencils	37	empty plastic bottles
16	broken pens	17	tattoo removal ads
1	pickle pen	14	ear buds
4	used Band-aids	1	coprolite
2	cell phones	2	shoes (unmatching)
1	Barney watch	142	squashed raisins
20	toenail clippings	1	copy of this book
1	master list of passwords for the central office computer	1	partridge in a pear tree

a. How many things did he find on the floor?
b. What was the ratio of hair ties to toenail clippings? *(reduce)*
c. What exact percent was the coprolite of the total garbage?

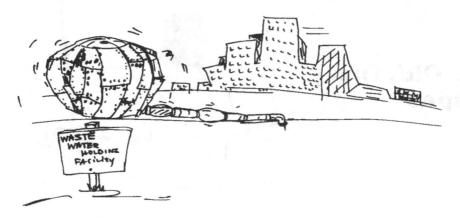

81. Overflow

FACT: The average toilet uses 1.5 gallons of water per flush.
If a city's wastewater treatment plant has three 100,000-gallon
capacity tanks for processing toilet water, how many toilets
flushing at once would it take to overflow the city's tank?

82. Cans of Spam™

**FACT: The average male in the US will consume
approximately 40 tons of food in his or her lifetime.** If Asa
ate and continues to eat nothing but 1/4-pound cans of Spam™
for his entire life, how many total cans of Spam™ would he eat
in his entire lifetime?

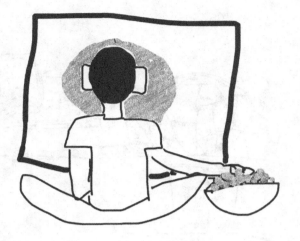

83. Old, fuzzy grapes

While playing a computer game, David was eating grapes out of a bag. The bag contained 60 grapes when it was full, and it is now 2/3 empty. David did not notice, but 14 percent of the grapes in the bag are old and fuzzy. How many old, fuzzy grapes has David eaten?

84. Frog eggs

FACT: Frogs lay 3,000 eggs at a time. You find yourself floating in a pond surrounded by eggs from 147 pairs of adult frogs. How many frogs will be surrounding you at the moment when all the eggs have hatched?

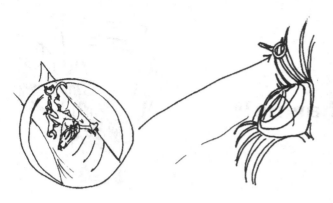

85. Eyelash mites

FACT: Your science book may not have told you this, but Demodex mites (which look like teeny tiny 8-legged crocodiles) live on everyone's eyelashes. These harmless mites slither down eyelashes into the eyelid where they lay their heart-shaped eggs. It takes one mite about 4 minutes to travel 20 mm. How long will it take a mite to slither down an eyelash 1 cm long?

86. Cow gas

FACT: Cows pass a lot of gas. A single cow burps out 600 pints of methane per day. Asa and David have been given permission to collect the methane from Asa's neighbor's cow, Blossom, over the course of one day. How many gallon jugs will they need to bring?

87. Lucky Charms™

Every morning beginning on March 28th, Asa pours a bowlful
of Lucky Charms™, scoops out all of the little blue moon
marshmallows from the bowl and puts them in a lockbox in his
closet. How many little blue moon marshmallows does Asa have
in his lockbox by the end of September, if the cereal averages
17 little blue moon marshmallows per bowl? (Asa insists we
include this problem.)

88. Cereal boxes

It is January 1st, and there are 6 boxes of cereal in David's
household. The wheat bran flakes will last 4 months (David's
mom is the only one who eats these), the Frosted Mini-
Wheats™ will last 3 1/2 weeks, the Rice Krispies™ will last
4 weeks, the Cheerios™ will last a week and a half, the Lucky
Charms™ only last 1 week, and the All Bran™ will never be
eaten by anybody. If all the boxes are presently full, how many
more boxes of each kind of cereal will David's mom have to buy
to make sure they don't run out of any cereal this year?

89. Back yard fertilizer

David's dog, Scruffy, "fertilized" the back yard in 400 evenly distributed places. The yard is 100′ long x 100′ wide. If tonight, while David searches for nightcrawlers, he steps on ten square feet of the yard, what percent chance is there of his shoes becoming smeared with dog dooky?

90. Taste buds

FACT: The average human has 9,000 taste buds. David's sister ate an entire bag of incredibly red-hot candies and temporarily numbed 47 percent of her taste buds in the process. How many taste buds are still functioning?

91. Grocery store bulletin board

Central Wisconsan Computer Club
"Learning to Use Spell Check"
We meat 8-9 p.m. Wendsday

FOR SALE Self-Driving School Bus
Low Mileage, Very few dents
700-444-1111

CROCODILE FOR SALE $1400
5 meters long.   Weighs 1100 kg
Requires pool of its own.
Bites may result in amputations.
Actually, whatever you want to pay
will be fine... or we'll pay you!
Call 489-489-333 PLEASE!

LAWN MOWING
I WILL NEED TO USE YOUR
LAWNMOWER AND I WILL
NEED TRANSPORTATION TO
YOUR PLACE. PAY ME $40 PER
HOUR INCLUDING TWO 15-
MINUTE REST BREAKS WITH
LEMONADE AND COOKIES.
EMAIL ME!

ROTTEN TOMATOES
That's the name of our band because
we are pretty bad. Still, we're
cheap! $25 per night or free!
715-477-1900

Time Travelers' Meeting
7:45 p.m. last Tuesday

TIME MANAGEMENT WORKSHOP
for people who find themselves just too busy.
We meeting 6:30-9:30 p.m.Mondays through Saturdays.
Some additional homework required.
Email Fred@toobsy.com

FOR SALE Self-Driving School Bus
Low Mileage, Very few dents
700-444-1111

Are you illiterate?
Just call 1-800-CAN'T READ
We understand what you're going through.

At the local grocery store bulletin board, Asa and David found
these 10 signs. Create three problems of your own based on
these signs.

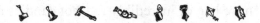

92. Leeches

FACT: Leeches suck up to eight times their weight in one sitting. If you were a 90-pound leech, how many 12-ounce cans of you-know-what could you suck up in one sitting? *(1 pound = 16 ounces)*

93. Royal Canadian Bagpipers

David's alarm clock is programmed to play the Royal Canadian Bagpipers at 7:08 every morning. If it takes David 9 1/2 minutes to wake up, reach over and push the snooze button, and 23 more minutes before he actually gets out of bed, how much time will he have to get ready before it's time to leave for school at 7:45?

94. Do you smell something?

FACT: Scruffy thought it would be fun to attack the fluffy black-and-white kitty that crawled under the porch. One spray of skunk can be smelled for more than a mile in any direction. How many square miles around David's house now stink with skunk stink? (And would anybody like to adopt Scruffy for a while?) *(Area of a circle = pi * radius²) (pi = 3.14)*

95. Tick eating habits

FACT: The average tick eats once every 7 years. The average American eats five times a day. How many times more often do Americans chow down than ticks? (*Use 365.25 for days in a year*.)

96. What would you do with $3,500?

FACT: Each year 1,000 cats in the United States have heart operations to receive pacemakers, at a cost of $3,500 each. What would you do with $3,500?

97. ASA

The entry fee for the After School Activities (ASA) Outdoor Mud Sculpture Contest was $0.25. Only 33 people entered and, because it started raining, no one was able to make anything worthy of a prize. Wanting to be fair, Asa divided the prize money ($5.61) equally among the entrants. How much did each one win, and what was Asa's profit?

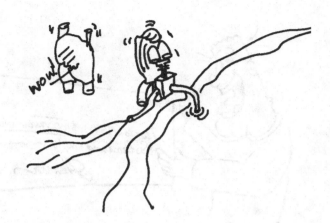

98. Glub

FACT: The heart pumps 2,000 gallons of blood each day. Lake Michigan contains 1.3 quadrillion gallons of water. How long would it take a pumping heart to fill something the size of Lake Michigan? *(Use 365.25 days in a year.)*

99. Mucus moves

FACT: Mucus moves through our body at the speed of 5 mm per minute. If you were riding in a tiny mucus-powered bus through your body, how long would it take you to go one meter? *(Show answer in hours and minutes.)*

100. Cellmates

FACT: A female rat can give birth six times a year, and a litter will contain about 10 little baby rats. Let's say you were imprisoned in a jail cell with one male rat (which you named Manfred) and one female rat (which you named Matilda). Matilda right away gave birth to 10 little baby rats and continued giving birth every two months. Half the babies born were male and half were female. All the baby rats began to reproduce after their first birthday and all their litters are half male and half female. How many rats would you have to keep you company by the end of the second year? (It might help to make a chart.)

101. Asa's airplanes

FACT: Asa has 4 planes hanging from his ceiling. The tacks holding them up are 1 cm long. Each tack is pushed in tightly; however the tacks are beginning to slide at the rate of 1.5 mm per day. How long in days/hours until the planes begin their final flight?

102. Up, up, and away

FACT: The average person passes about a pint of gas daily.
How long would it take a person to fill a 5,000-gallon hot air
balloon with his or her own body gas? *(2 pints = 1 quart)*
(4 quarts = 1 gallon) (365.25 days = 1 year)

103. Floss 'em

Asa is supposed to floss his teeth using 18 inches of floss daily.
He began a new 50′ roll on March 17th. If he still has 2 feet of
floss left after flossing on August 2nd, how many days has he
forgotten to floss?

104. Saliva mouth

FACT: The average mouth produces about 2 ½ pints of saliva a day. Asa tried to test this hypothesis, and he did not swallow his saliva at all, but let it all drool into a 4-gallon bucket instead. How long before the bucket was full? You may wish to try this experiment at home. (Just kidding.)

105. What's that in my water bottle?

FACT: A full bladder has a radius of two inches. Would a perfectly round, full bladder fit into a water bottle with a circumference of 12"? (Show your work.)

106. Breathe in

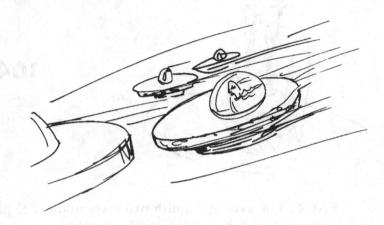

FACT: With each breath, we take in about a third of a cubic foot of air. There are approximately 4 million bacteria in each cubic yard of air—most of them important to your health. Inhale. How many bacteria just entered your lungs?

107. Arteries

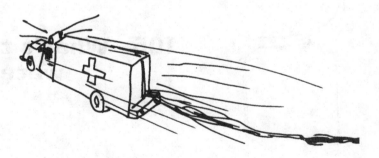

FACT: There are 2,451 air miles between New York and Los Angeles. There are approximately 62,000 miles of blood vessels in your body. If you stood in New York and stretched your blood vessels all the way to Los Angeles and then back again to New York and then to Los Angeles and so on, would you finish closer to New York or to Los Angeles? *(Show your work.)*

108. Live with Asa!

Asa is planning his own talk show, and has determined that a guest must have one trait from each of the categories below:

Descriptor	Occupation	Specialty
yodeling	skateboarders	who live in the Arctic
hula dancing	birdwatchers	who sing opera
barefoot	chefs	with super-long fingernails
double jointed	coal miners	who always walk backwards
billionaire	unicyclists	who are grandmas
ordinary	normal people	without any problems

How many combinations are possible? Which combination do you like best? (Unfortunately, as of the date of this publication, Asa has not been able to locate anyone fitting the last three.)

109. A rat can fall

FACT: A rat can fall 50 feet and land on its feet uninjured.
If a scientist (accidentally) dropped a rat out of the fifth-floor
window of a building where the first floor is on the ground
and each floor is 11 feet above the one below it, would the rat
survive? *(Show your work, please.)*

110. Censorship!

**FACT: David's mom censored 7.5 percent of Asa and
David's ideas for problems in this book.** How many problems
will you never see?

111. Smork

David and Asa have entered the "Say nothing but Smork" challenge and are doing great. Well, not really. The truth is Asa yelled at his sister for barging into the bathroom just 45 minutes into the contest. And David had to quit after an hour and a quarter because his mom refused to take him to school unless he said something besides "Smork." What is the ratio of Asa's Smork time to David's Smork time? *(Simplify.)*

112. Beetle in the soup

FACT: Beetles account for roughly 20% of all species—plant, animal, bacteria and fungi—known to science. There are about 400,000 different species of beetles. How many total species are known to science? (And what's that swimming in your soup?)

113. Good to know

FACT: An average American adult produces about 200 grams of feces per day. How many pounds of poop does the average American adult produce per day? *(Round to the nearest tenth.) (1 pound = 453.6 grams)*

114. Sneeze

FACT: Air from a big sneeze can travel 104 miles an hour. Asa is about to sneeze, and it's going to be a big one. How far away should David stand if the sneeze particles will be airborne for 2½ seconds? *(5,280 feet = 1 mile)*

115. Giant squid eyeball

FACT: The diameter of a basketball is about 9″. The diameter of a giant squid eyeball is 15″. If you were playing basketball with a giant squid eyeball and totally deflated it (through no fault of your own), how many cubic inches of giant squid eyeball juice would seep out onto the floor? *(The formula to find the volume of a sphere is $(V = 4/3 \cdot \pi \cdot r^3)$.)*

116. Did I tell you my birthday is July 31st?

If David mentions his birthday every day throughout the month of July—once on July 1st, three times on July 2nd, five times on July 3rd, and so on for the entire month—how many times will he mention his birthday on July 30th? (By the way, his birthday is on July 31st. David says cards and letters may be sent to him at the address printed in the front of the book. Presents will also be accepted. Contact him for gift ideas first.)

117. Belly button biodiversity

FACT: Scientists swabbed the navels of 60 random Americans to see what microbes were lurking there. The study found 2,368 species of bacteria. Of those, only 910 had been previously known to science. How many new-to-science microbes were found per belly button? (And what interesting microbes are living in yours?)

118. Heavy thinker

FACT: The average brain weighs three pounds. After doing this math book, the average brain will weigh about 0.5 pounds more. If someone completes the whole book, what percent larger will their brain be? *(Round to the nearest whole percent.)*

119. In the Aftermath Restaurant

David's mom, David, and Asa went to this restaurant to eat.
How much did David's mom pay for the three of them?

Asa's order:
Macaroni and Sneeze
Large Classic Choke with cup
Rhubarb Wire Pie—4 feet.

David's mom's order:
Deep fried lint with French flies
Sweep potato pie
Water

David's order
Thinsulate Pancakes
Large Arti Choke with cup
Last Stand Custard

AFTERMATH RESTAURANT MENU	
Entrees:	
Macaroni and Sneeze	$15.20
Pseudo-Chicken Parts in white whine sauce	$14.25
Deep Fried Lint (please specify dryer or pocket)	$17.25
Thinsulate Pancakes in clogged cabin syrup	$14.68
Wax Fruit Bowl delicately scented with smelly markers	$16.21
Value Meal: photocopied foods	$ 0.10 per sheet
...any order with French flies	$4.50 additional
Beverages, any size	$1.50
Classic Choke, Cherry Choke, Chokecherry, Diet Choke, Artichoke, Pure Caffeine	with cup: add $1.00
Desserts:	
Last Stand Custard	$ 7.25
Baked Alaska (actual chunks included)	$ 8.95
Ice Scream	$ 8.49
Rhubarb Wire Pie, per foot	$ 3.00
Sweep Potato Pie	$10.55
Hot Sludge Sundae	$14.60
	add 5% sales tax
Eat at your own risk Management accepts no responsibility for health of its customers.	

120. Care to join us?

We invited you to join us at the Aftermath, where there are so many different combinations of delectable yummies! How many different combinations (not to mention indigestion) would be available if you chose one entrée (without flies), one beverage (no cup) and one dessert? ... and what's your excuse for not coming?!?

Answer key

Warning: answers may vary due to rounding.

That's cool.

1. Please don't

If your skin is laid out in a square, what's its length and width?

25 square feet = **5 feet long, 5 feet wide**

2. Slosh

How many cups of water are in the brain?

1 gallon water = 8 pounds

brain water = 2/8 = 1/4 gallon

16 cups = 1 gallon

1/4 of 16 = **4 cups of water in the brain**

3. What's all that racket!

Approximately how many miles high do they fly?

32,000 feet ÷ 5,280 feet = **approximately 6 miles**

4. Ice pops

How many degrees different are the Siberian salamander and David's sister?

68 – (−58) = 68 + 58 = **126 degrees**

5. I'll be with you in a jiffy

How many milliseconds?

1 second = 1,000 milliseconds

12 seconds = **12,000 milliseconds**

6. Doggie menu

How many calories will Scruffy still need to consume today?

1500 − (325 + 20 + 50 + 170 + 40 + 1) = **894 calories**

7. It wasn't me!

Find the mean number of times a person passes gas daily.

(14 + 23) ÷ 2 = **18.5 times a day**

8. Crayfish smoothie!

How many more crayfish should Asa eat than his twin sister (to the nearest tenth)?

Boys: 2800 ÷ 70 = 40 crayfish

Girls: 2200 ÷ 70 = 31.4 crayfish

40.0 − 31.4 = **8.6 crayfish**

9. Popcorn gone bad

How many tons did the popcorn ball weigh?

3423 pounds ÷ 2000 pounds = **1.7115 tons or 1 ton 1,423 pounds**

10. Crumbly

How many crumbs are still on the floor?

$17 * 3 = 51$

$51 - 42 =$ **9 crumbs**

11. Shrimp, huh?

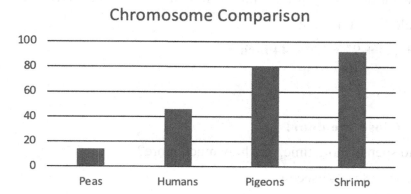

Chromosome Comparison

12. PFFFFFFT

Find the average amount of gas expelled per poot.

$500 + 1500 = 2000$

$2000 \div 2 = 1000$ ml gas per day

1000 ml $\div 16.5$ poots/day = **60.6 ml per poot**

13. Dandelion doughnuts

What percent of edible plant species aren't being eaten?

$80,000$ edible $- 30$ eaten $= 79,970$ not eaten

$79,970 \div 80,000 = 0.999625$

$0.999625 =$ **99.96% of edible plants are not being eaten regularly**

EXTENSION: Make a list: How many different plant species do you eat?

14. I'm coming as a dung beetle

How many of the 600 people should come dressed as insects?

80% of 600 = 0.80 * 600 = **480 really strange people**

15. Fly tracks

How far did the fly walk? Not far. It died. But had it survived, it would have traveled:

9″ + 13″ + 9″ + 13″ = **44 inches**

16. Close the door!

Who spends more time, and how much more?

Asa 18 * 8 = 144 seconds

David 12 * 12 = 144 seconds

They both spend equal time.

17. Stop staring!

How many mites might be on this paper now?

8.5 inches * 11 inches = 93.5 square inches

93.5 * 5 million = **467.5 million or 467,500,000 mites**

18. Bring in the SWAT team

335,923,200,000,000 * 12 swats/baby = 4,031,078,400,000,000 swats 4,031,078,400,000,000 swats + 24 swats for the original flies who caused the overpopulation problem in the first place = **4,031,078,400,000,024 swats**

19. Care to go for a swim?

How many cups of spit would fill one swimming pool?

(4 cups = 1 quart)

$25,000 \div 2 = 12,500$ quarts

$12,500$ quarts $* 4$ cups/quart = **50,000 cups of spit**

20. Star pupil

How many minutes and seconds until all 21 dogs were howling?

10 seconds $* 21 = 210$ seconds

210 seconds $\div 60$ sec/min = **3 minutes 30 seconds**

21. Cheese!

How long does it take Scruffy to come for peanut butter?

$0.84 + 0.25 =$ **1.09 seconds**

22. Lawn ornaments

Determine the sum: **120%**

How is that possible?

Some houses have more than one lawn ornament.

23. It's normal

How many times would this normally occur in math class?

15 times/day $* 28$ people = 420 times/day

420 times/day $\div 24$ hours/day = **17.5 times/hour**

24. But it was good dirt

How many grams of dirt have you ingested in the last ten years?

100 milligrams ÷ 1000 milligram/gram = 0.1 gram/day

0.1 gram/day * 365.25 days/year = 36.525 grams/year

36.525 grams/year * 10 years = **365.25 grams of dirt**

25. Stop the car!

60 minutes/hour * 3.5 hours = 210 minutes

210 minutes * 2 ml/minute = 410 ml full

No! He will not make it! (Mom!! Mom!!!)

26. Let me in!

54 sec * 17 mosquitoes/sec = **918 mosquito bites**

27. Drowning in droplets

How many saliva droplets expelled during lecture?

300 saliva droplets * 55 minutes = **16,500 saliva droplets**

28. Eliminate waste

How many inches per hour does waste move through the colon?

1 1/2 day = 24 + 12 = 36 hours

36 inches in 36 hours = **1 inch per hour**

29. Pooky

How many hours and minutes does Pooky have to do other cat things?

85% of 24 hours = 0.85 * 24 = 20.4 hours

24 − 20.4 = 3.6 hours

0.6 * 60 = 36 minutes

3 hours 36 minutes

30. Bats in the tent

How many little brown bats to invite?

135,000 mosquitoes ÷ 500 per bat = **270 bats** (or one superbat)

31. Flapflapflapflap

How many times faster is the bee?

1 second: 2 minutes 20 seconds =

1 second: 140 seconds = **140 times faster**

32. Thank you, garbage people

How much did each one's trash weigh after three days?

4.4 * 3 = **13.2 pounds of trash**

EXTENSION: About the same as 1 gallon of paint, 2 quarts of juice or a bowling ball. *(Answers will vary.)*

33. Did you remember to flush?

How much has the Space Shuttle toilet cost per flush?

$23,400,000 dollars ÷ 121,875 times used = **$192.00 per flush**

34. Slimy

How many hair scrunchies will he have found?

21 hair crunchies / 7 weeks = 3 hair scrunchies per week

21 hair scrunchies + 3 hair scrunchies = **24 scrunchies**

(Hair thingies for sale cheap! Variety of colors. Contact Asa.)

35. Tern, tern, tern

How many frequent flier miles could an Arctic tern collect over 20 years?

11000 * 2 * 20 = **440,000 frequent flier miles**

36. Zit City

How many zits if every oil gland breaks forth into a zit?

200 million hairs × 2.5 oil glands/hair = **500 million zits**

37. Look closely

A 2-gallon vacuum bag filled with dust contains how many little mites?

1 gallon = 16 cups, 2 gallons = 32 cups

32 cups * 16 tablespoons/cup = 512 tablespoons

512 tablespoons * 3 teaspoons/tablespoon = 1536 teaspoons dust in bag

1/2 teaspoon contains 1000 mites, so 1 teaspoon contains 2000 mites

1536 teaspoons * 2000 mites/teaspoon = **3,072,000 mites**

38. Worm digging

How many worms can they expect to dig up on Asa's 3/4 acre front lawn?

3/14 of 3,000,000 = 0.75 * 3,000,000 = **2,250,000 worms**

39. Panda poop

How many pounds of poop does a panda poop per day?

4/5 of 30 =

4/5 * 30 = 120/5 = **24 pounds of poop**

40. Ants rule

How many billion ants?

20,000,000,000,000,000 = 20 quadrillion

20,000 * 1,000,000,000 = 20,000 trillion

20,000,000 * 1,000,000,000 = **20 million billion ants**

41. Hot dog

How high would a hotdog bounce from:

    a. the top of Asa's garage

       13 feet * 0.12 = **1.56 feet**

    b. the top of City Hall

       85 feet * 0.12 = **10.2 feet**

    c. a Boeing 747

       12,000 meters * 0.12 = **1,440 meters** or **1.44 kilometer**

42. Snarly

How many tangles wil Scruffy have 181 days later (to the nearest hundredth)?

15 new tangles − 14.25 removed per day = 0.25 more per day

0.25 ∗ 181 = **45.25 tangles**

43. Supersweeper

What fraction of the junk has it collected?

100/100 million = 1/1 million

1/1 million or **1/1,000,000**

44. I don't get it

How many saliva droplets released?

2.5 saliva droplets/word ∗ 4 words = **10 saliva droplets**

45. Where did my mailbox go?

How many mailboxes will she have hit?

2 1/2 hours driving ∗ 60 minutes per hour = 150 minutes

150 minutes ÷ 15 minutes per mailbox = **10 mailboxes**

46. Skin scales 25 cents a pound (Do I hear 30!?)

How much could David make selling his skin scales?

16 ounces per pound ∗ $0.25 per ounce = **$4.00 per year**

47. Smile!

How long does it take a rattlesnake to slither 880 yards?

880 yards * 3 feet/yard = 2640 feet

2640 feet ÷ 5280 feet/mile = 0.5 miles

0.5 mile ÷ 2 miles/hour = 0.25 hour

0.25 hour * 60 minutes/hour = **15 minutes**

48. I thought I smelled something

How many school days later did he discard the fish?

147 ÷ 3.5 = **42 school days later**

49. Coffee, anyone?

How many 8-ounce coffee mugs of sweat?

8 ounces = 1 cup

2 cups = 1 pint

1 1/2 pints = 2 + 1 = 3 cups = **3 mugs**

50. A very hairy ordeal

How long would your hair be in 15 years?

0.35 millimeters * 365.25 days/year * 15 years = 1917.5625 millimeters

1917.5625 millimeters ÷ 1000 millimeters/meter = **1.9 meters long**

Send us your scale drawing. We'll send you a lock of hair.

51. Locker room

What will be the average number of sweat glands going into the locker room?

$(2{,}000{,}000 + 5{,}000{,}000) \div 2 = 3{,}500{,}000$ sweat glands average per person $3{,}500{,}000 * 35$ people = **122,500,000 sweat glands** (17% dripping profusely)

52. Nematode sea worms

How many kilometers long would the nematode line be? 40, 000,000,000,000,000,000,000,000,000 $* 2$ mm $* 1m/1000mm *$ $1k/1000m =$ **80,000,000,000,000,000,000 kilometers**

53. Adventure awaits

For example, 15 miles round trip $*$ 180 days of school = 2,700 miles—**Guatemala!** *(Answers will vary. If they don't vary, somebody's peeking.)*

54. Classical Kazoo Symphony

How much of the Kazoo Orchestra was he subjected to?

4 minutes $*$ 14 boxes of junk = **56 minutes**

55. Mirror, mirror, on the wall

How much time does she spend every day in front of the mirror?

14% of (24 hours $-$ 9 hours sleeping) =

14% of (15 hours) = $0.14 * 15 = 2.1$ hours per day

$0.1 * 60 = 6$ minutes

2 hours 6 minutes

56. After School Science

Hours	Minutes	Seconds
5	0	45
3	17	30
0	86	00

8 hours, 103 minutes, 75 seconds

103 minutes = 1 hour 43 minutes

75 seconds = 1 minute 15 seconds

(8 + 1 hours) + (43 minutes + 1 minutes) + (15 seconds)

9 hours, **44 minutes, 15 seconds**

57. Human vs. pineapple

What percent fewer chromosomes do humans have than pineapples?

25 − 23 = 2 fewer chromosomes

2/25 = 0.08 = **8% fewer chromosomes than pineapples**

58. You're next, young man

How many total millimeters did he bite off?

1.2 cm ∗ 10 mm/cm = 12 mm

12 mm ∗ 0.05 = 0.6 mm from each

0.6 mm ∗ 10 fingers = **6 mm total**

59. How many didn't bite?

25 million assorted insects over every square mile

10 square miles ∗ 25 million insects = 250 million insects over Camp Torture

Asa was bitten by 1,738 insects.

250,000,000 − 1,738 = **249,998,262 insects did not bite Asa**

60. Great grandma's knitting

Now how much more does she have to knit?

1/2 of 3/4 of complete project became unraveled (Hint to students: "of" means "times".)

1/2 * 3/4 = 3/8 of complete project became unraveled and 3/8 is still intact

1 – 3/8 = 5/8 **more to do**

61. Whole-wheat rutabaga caramel tofu surprise

How many slices did they each eat?

Asa ate 39 chunks more ÷ 13 chunks/slice = Asa ate 3 slices more than David

If X stands for Asa's slices, then X + (X−3) = 11

2X − 3 = 11

2X = 14

X = 7

David ate 4, and Asa ate 7.

62. Fungus among us

How much of the park will be covered?

To find the area of the park, multiply length * width

You can use fractions:

1 1/2 * 1 1/2 = 3/2 * 3/2 = 9/4 = 2 1/4 sq mi

Or use decimals:

1.5 * 1.5 = 2.25 sq mi

Either way, **the whole park would be covered.**

63. Fuzzy apple logic

What percent of the original apple will be left?

0.04 ∗ 14 days = 0.56 = 56% of its mass it has lost

100% − 56% = **44% of its original mass is left**

64. Let us see your hair, David

How many days will he have to wear his MLOS hat after the next haircut?

last haircut	Oyster Shuckers Bowling League hat	3 days
one before	Mime Radio Fan Club cap	4½ days
before that one	Camp Torture hat	6 days

Fortunately, his mom must be improving (or else David is getting used to funny hair). Each day the amount decreases by 1½ days. At this rate he will only have to wear his Marshfield Lawn Ornament Society hat **1½ days.**

65. Shhhhh

How many ideas does Asa lose each day?

320 ideas ∗ 0.20 = 64 ideas interrupted

¼ ∗ 64 = 16 return later

64 lost − 16 which return = **48 lost per day**

(And the world is a better place because of it.)

66. N.L.R.M.I.T.

What percent of refrigerators contain more than one n.l.r.m.i.t.?

7 out of 8 contain more than one n.l.r.m.i.t

7 out of 8 = 7/8 = 7 ÷ 8 = 0.875

To find part of something, multiply the part times the something.

0.875 ∗ 0.70 = 0.6125 =

61.25% of refrigerators have more than one n.l.r.m.i.t.

67. Sludge

How deep is the sludge pool?

length × width × depth = volume

25 ft × 50 ft × depth = 157,000 cu ft

1,250 sq ft × depth = 157,000 cu ft

157,000 cu ft ÷ 1,250 sq ft = **125.6 feet deep**

68. No, we are not buying a 3D TV

For each price add the $40 delivery fee. Then you could multiply by 0.05 and add the price, but save a step and multiply ∗ 1.05. Then round to the nearest cent.

- Genuine 100% cathair couch

  1,039.99 + $40 = $1079.99

  1,079.99 ∗ 1.05 = **$1133.99**

- Overstuffed Quicksand Couch

  1,412.99 + 40 = 1,452.99

  1,452.99 ∗ 1.05 = **$1,525.64**

- Giant Whoopee Cushion Bean Bag Chair

  1,123.58 + 40 = 1,163.58

  1,163.58 ∗ 1.05 = **$1,221.76**

- Prestained Wonder Couch

  1,421.06 + 40 = 1461.06

  1,461.06 ∗ 1.05 = **$1,534.11**

- Lazy Person Couch

  1,475,380.64 + 40 = 1,475,420.64

  1,475,420.64 ∗ 1.05 = **$1,549,191.67**

69. Camp Torture

How many total sandwiches had they found?

3/8 = 3/8 egged salad sandwich

1/4 = 2/8 artificial preservative spread

3/8 = 3/8 llama milk cheese on Catcher in the Rye bread

5/8 = 5/8 knuckle sandwich

7/8 = 7/8 toe-food and toe-mato on hole wheat crisps

1/2 = 4/8 Salmon Ella sandwich

3/4 = 6/8 lox and keys on unleveled bread

1/4 = 2/8 Soilent Green Burger Deluxe

Total = 32/8 = 4 whole sandwiches

70. Heinz and Gunter

What part of each type of sandwich was now left?

3/8 – 2/8	= 1/8 egged salad sandwich
1/4 – 1/4	= 0 artificial preservative spread
3/8 – 2/8	= 1/8 llama milk cheese on Catchup in the Rye bread
5/8 – 2/8	= 3/8 knuckle sandwich
7/8 – 2/8	= 5/8 toe-food and toe-mato on hole wheat crisps
1/2 – 1/4	= 1/4 Salmon Ella sandwich
3/4 – 1/4	= 1/2 lox and keys on unleveled bread
1/4 – 1/4	= 0 Soilent Green burger deluxe

71. Gum on the seats

How many seats in Yankee Stadium have gum stuck to them?

2 out of 3 = 2/3 have gum

1/3 of 52,326 = 52,326 ÷ 3 = 17,442

Therefore 2/3 of 52,326 = 17,442 * 2 = **34,884 gum-stuck seats**

72. Chicken noodle sloop

How many skin scales containing large colonies of bacteria just fell into the soup?

750 * 2/3 = 750 * 2 ÷ 3 = **500 large colonies of bacteria**

73. Candy jar

How many revolting butterscotch candies?

2/3 of 12 = 2/3 * 12 = (2 * 12) ÷ 3 = **8 revolting butterscotch candies**

74. People shed too

How many skin scales can you expect to peel off your body in the next 20 minutes?

10 billion/day ÷ 24 hours/day = 0.4166666 billion per hour

20 minutes/60 minutes = 1/3 of an hour

0.41666666 billion ÷ 3 = **0.1388888 billion skin scales**

or 0.1388888 billion * 1,000,000,000 = **138,888,800 skin scales**

(Answers may vary somewhat due to rounding differences)

75. Lumps

What percent of the gravy would not go through the strainer?

48 ounces total − 26 ounces which went through = 22 ounces did not

22/48 = 22 ÷ 48 = 0.458333 = 45.8% = **46% of gravy**

76. Taj Mahal

How many cubic-foot bricks of snow to replicate the Taj Mahal?

cubic feet = length * width * depth

$110 * 50 * 2 =$ **110,000 bricks**

77. Longest toe

Use a ruler, graph paper or a computer to create a bar graph showing the number of people with each longest toe.

78. An acre of pizza, please

How many whole pizzas are eaten per half hour?

350 slices/second * 60 sec/min * 30 min/half hour = 630,000 slices per half hour

630,000 slices ÷ 8 slices per pizza = **78,750 pizzas/half hour**

(Would you like that delivered, or do you plan to eat it here?)

79. Pair of socks

How many grams heavier will his socks be?

$2 * 365 = 730$ days straight wearing

$730 * 190$ mg $= 138,700$ mg heavier $=$ **138.7 grams heavier**

(Usually every 6–8 weeks he empties them out.)

80. School janitor

a. How many things did he find on the floor? = 425

b. What was the ratio of hair ties to toenail clippings? *(reduce)*

Ratio hair ties to toenail clippings = 84:20 or 84/20 = 21/5

c. Fraction coprolite of total = 1/425 = 1 ÷ 425 = 0.002352941176471

0.002352941176471 * 100 = **0.2352941176471% of total finds**

81. Overflow

How many flushes to overflow?

1.5 gallons per flush

City can handle 3 * 100,000 gallons = 300,000 gallons

300,000 gallons ÷ 1.5 gallons/flush = 200,000 toilets flushing is

the maximum 200,000 +1 = **200,001 toilets flushing in unison**

to overflow

(Most likely to happen during Superbowl halftime.)

82. Cans of Spam

How many 1/4-pound cans of Spam?

40 tons * 2,000 pounds per ton = 80,000 pounds of Spam™

eaten

80,000 pounds ÷ 1/4-pound cans = **320,000 cans of Spam™**

83. Old, fuzzy grapes

How many old, fuzzy grapes has David eaten?

David has eaten 2/3 of 60 = 2/3 x 60 = 40 grapes

14% of 40 grapes = **5.6 grapes**

(He's still chewing the sixth moldy, fuzzy grape.)

84. Frog eggs

How many frogs will be surrounding you when the eggs have all hatched?

Maybe you thought: 147 * 3,000 = 441,000 baby frogs

441,000 baby frogs + 147 male frogs + 147 female frogs =

441,294 frogs

Actually, frog eggs hatch into tadpoles. The only frogs surrounding you now are the same 147 pairs that were there when you got in.

males + 147 females = **294 frogs**

(But unless you want to get into a debate, be kind and accept both answers.)

85. Eyelash mites

How long to slither one centimeter?

centi- means *hundred* (100), *milli-* means *thousand* (1,000)

Therefore 1 cm = 10 mm

20 mm in 4 minutes

10 mm in **2 minutes**

Or you could write an equation:

4 minutes/20 mm = x minutes/10 mm

cross multiply: 40 = 20x

x = **2 minutes**

86. Cow gas

How many gallon jugs of cow gas?

2 pints/quart * 4 quarts/gallon = 8 pints per gallon

600 pints/8 pints per gallon = **75 gallon jugs**

87. Lucky Charms™

How many little blue moon marshmallows by the end of September?

March 28–March 31	4 days
April	30 days
May	31 days
June	30 days
July	31 days
August	31 days
September	30 days

Total: 187 days of putting little blue moon marshmallows in his closet

187 days ∗ 17 little blue marshmallows per bowl = **3,179**

marshmallows

88. Cereal boxes

How many more boxes to buy before the end of the year?

Wheat bran flakes: 12 months ÷ 4 months = 3 boxes/year

They will need 3 boxes for 12 months, but they already have one,

so 3 − the one they already have = **2 boxes Wheat Bran flakes**

Frosted Mini-Wheats™: 52 weeks ÷ 3.5 weeks = 14.85 boxes = 15 boxes/year

15 boxes − the one they already have = **14 boxes Frosted Mini-Wheats™**

Rice Krispies™: 52 weeks ÷ 4 weeks = 13 boxes /year

13 boxes − the one they have = **12 boxes Rice Krispies™**

Cheerios™: 52 ÷ 1.5 weeks = 34.6666 = 35 boxes/year

35 boxes − the one they have = **34 boxes Cheerios™**

Lucky Charms™: 52 ÷ 1 week = 52 boxes/year

52 boxes − the one they have = **51 boxes Lucky Charms™**

All Bran: 1 box − the one they have = **0 boxes All Bran™**

89. Back yard fertilizer

What is the percent chance of David's shoes being smeared with dog dooky?

100′ x 100′ = 10,000 square feet total

He walks on 10 square feet out of 10,000 sq. feet = 1/1,000 of total lawn area

If he walks on the whole yard, 100% of the time he will step on 400 droppings.

Walking on 1/1,000 of the yard, his odds are 1/1000 * 400 = 40/100

40/100 = **40% chance** (David beat the odds.)

90. Taste buds

How many taste buds are still functioning?

9,000 taste buds * 47% numbed

9,000 * 0.47 = 4,230 numbed

9,000 − 4230 = **4,770 taste buds still functioning**

91. Grocery store bulletin board

It says, "three problems of your own." You want us to do all the work?

92. Leeches

How many cans of you-know-what would you suck up?

90 pounds * 8 = 720 pounds of you-know-what

1 pound = 16 ounces

720 pounds * 16 ounces/pound = 11,520 ounces

11520 ounces ÷ 12 ounces in a can = **960 cans of you-know-what**

93. Royal Canadian Bagpipers

How long will he have to get ready?

7:08 a.m. + 9 1/2 min + 23 min = 7:40 1/2 when he finally gets out of bed

7:45 a.m. − 7:40 1/2 = **4 1/2 minutes to get ready**

94. Do you smell something?

How many square miles around David's house smell like skunk?

$a = \Pi r^2$

area = 3.14 * (1 * 1)

area = **3.14 square miles** or more specifically, **3.1415926535897 93238462643383279502884197169399375... square miles**

95. Tick eating habits

How many times more often do Americans eat than ticks?

7 years compared to 1/5 day

365.25 days * 7 = 2,556.75 days

2,556.75 days compared to 1/5 of a day

2556.75 / 1/5 = 2556.75 * 5 = **12,783.75 times more often than ticks**

96. What would you do with $3,500?

Asa and David would donate the greater portion of the money to their ongoing study of interspecies communication. The rest to buy burritos.

97. ASA

How much did each one win, and what was Asa's profit?

5.61 ÷ 33 people = **$0.17 each won**

33 people * 0.25 = $ 8.25 entry fees paid

$8.25 − 5.61 = **$2.64 Asa's profit**

98. Glub

How long would it take a heart to fill something the size of Lake Michigan? 1,300,000,000,000,000 gallons ÷ 2,000 gallons per day = 1,300,000,000,000 gallons ÷ 2 = 650,000,000,000 days

650 billion days ÷ 365.25 days/year = 1.779603 billion years

It would take **1,779,603,000 years**.

You may begin.

99. Mucus moves

How long would it take the mucus bus to go one meter?

1 meter = 1000 mm

1,000 mm ÷ 5 mm/min = 200 minutes

200 min ÷ 60 min/hr = 3.33333 hours = 3 1/3 hours

1/3 * 60 = 20 minutes

3 hours 20 minutes

100. Cellmates

How many rat friends by the end of the second year?

Step #1. Start with 2 original rats, Manfred and Matilda.

Step #2. First year: Matilda gives birth 6 times * 10 rats/litter = 60 baby rats

Step # 3. Second year: Matilda gives birth 6 times * 10 rats/litter = 60 more rats

Step # 4. Also, the 60 rats born first year reproduce, each female gives birth once. 30 females * 10 babies each = 300 new rats

Total: 2 + 60 + 60 + 300 = **422 rats to keep you company**

101. Asa's airplanes

How long until tacks and planes fall?

1 centimeter = 10 millimeters

10 mm ÷1.5 mm per day = 6.6666 = 6 2/3 days

2/3 of 24 hours = 2/3 * 24 hours = 16 hours

6 days, 16 hours

102. Up, up, and away

How long to fill a hot air balloon?

1 gallon = 4 quarts * 2 pints each = 8 pints

5,000 gallons * 8 pints/gallon = 40,000 pints in a hot air balloon

40,000 pints ÷ 365.25 days/yr = **109.5 years** (or about one long lifetime)

103. Floss 'em

How many days has Asa forgotten to floss?

March 17th–March 31st = 15 days (count both the beginning and the end days)

15 days March + 30 April +31 May + 30 June + 31 July +
2 August = 139 total days

50′ roll − 2 feet of floss left = 48 feet used

18 inches supposed to floss per day = 1½ feet per day

48 ft ÷ 1 ½ ft/day = 32 days he actually flossed

139 days − 32 days =**107 days he forgot**

104. Saliva mouth

How long before the bucket was full?

2 pts/qt ∗ 4 qt/gal = 8 pt/gal

8 pt/gal ∗ 4 gal/bucket = 32 pt/bucket

32 pt/bucket ÷ 2.5 pints/day = **12.8 days** or...

12.8 days = 12 days + 0.8 day = 0.8 day ∗ 24 hr/day =
19.2 hours

0.2 hr ∗ 60 min/hr = 12 minutes or **12 days, 19 hours, 12 minutes**

105. What's that in my water bottle?

Would a full bladder fit?

Radius = 1/2 diameter, therefore the bladder diameter = 4 inches

circumference = 3.14 ∗ diameter

circumference of bladder = 3.14 ∗ 4 inches = 12.56 inches

Nope!

106. Breathe in

How many bacteria just entered your lungs?

1 cubic yard = 3 ft ∗ 3 ft ∗ 3 ft = 27 cubic feet

4,000,000 per cubic yard ÷ 27 = 148,148 bacteria/cubic foot

1/3 cubic foot ∗ 148,148 bacteria =

49,383 bacteria (most of them very nice!)

107. Arteries

Would you finish closer to NY or LA?

62,000 miles of blood vessels ÷ 2,451 miles = 25.295797 trips

Your first trip (and all odd-numbered trips) will end in LA.

Your second trip (and all even-numbered trips) will end
in NY.

Therefore, 25.295797 trips would end **closer to LA**. (unless you
get lost and end up in Marshfield, Wisconsin…in which case,
hey, look us up!)

108. Live with Asa

How many combinations are possible? Which combination do
you like best?

There are 6 choices on each list. We may use each choice more
than once. Therefore, 6 ∗ 6 ∗ 6 = **216 possibilities**

109. A rat can fall

Floor of first floor = 0 feet

2nd-floor floor = 11 feet

3rd-floor floor = 22 feet

4th-floor floor = 33 feet

5th-floor floor = 44 feet

If the scientist was leaning out a window on the fifth floor, the rat would be just a few feet above the floor, and less than halfway to the ceiling.

11 feet × 4.5 high = (less than) 49.5 feet

The rat would survive.

110. Censorship

Let T = total number of problems before censoring

T = 107.5% of 120

T = 1.075 ∗ 120 = 129

129 − 120 = **9 censored problems**

111. Smork

What is the ratio of Asa's Smork time to David's? *(Simplify.)*

1 1/4 hours = (60 ∗ 1.25) = 75 minutes

Asa 45: David 75

Divide 45 and 75 by their common factors of 5 and 3 =

Asa 3 : David 5

112. Beetle in the soup

How many species are known to science?

x = species known to science

400,000 = 18% of x

400,000 = 0.18 * x

x = 400,000/0.18

x = 2,222,222 species total

113. Good to know

How many pounds are pooped? *(Round to the nearest tenth.)*

1 pound/453.6 grams * 200 grams/day =

200 grams/day ÷ 453.6 grams/pound = 0.4404 = **0.4 pounds**

114. Sneeze

How far away should David stand? What should I multiply? What should I divide?

Here's a tip to make any complex problem simple:

Start with the word you are looking for (in this case, *feet*) in the numerator: 5280 ft/1 mi

Then position what you know next to it, placing the words on the top or on the bottom so the words will cancel out.

Like this: 5280 ft/1 mi * 104 mi/hr * 1 hr/60 min * 1 min/60 sec * 2½ sec

Miles is on top (numerator) and on the bottom (denominator), so the words cancel out. Likewise, *hour*, *minutes*, and *seconds* cancel each other out. Only the word *feet* remains, which is what you want.

Now solve the equation, using your friend the calculator, and ask yourself:

Does my answer make sense? 381.33333 rounded = **381.3 feet away**

115. Giant squid eyeball

How many cubic inches of giant squid eyeball juice?

volume of sphere = $4/3 * \pi * r^3$

radius = 1/2 = diameter = $15 \div 2 = 7.5$ inches

vol = $4/3 * 3.14 * (7.5 * 7.5 * 7.5)$

vol = $1.33 * 3.14 * (7.5 * 7.5 * 7.5)$

vol = $1.33 * 3.14 * 421.875$

vol = **1,761.83 cubic inches of eyeball juice**

116. Did I tell you my birthday is July 31st?

How many times will he mention his birthday on July 30th?

Look for a pattern:

    July 1st—1 time

    July 2nd—3 times

    July 3rd—5 times

    July 4th—7 times

An astute mathematical wizard like yourself can see the relationship between the date and the number of times that David mentions his birthday.

It's 2 times the date minus 1.

Use this formula $(2 * d - 1)$ to determine the number on July 30th.

$2 * 30 - 1 = $ **59 times**

EXTENSION: So you want extra credit, do you? How many total times did David mention his birthday throughout the month of July, up to and including July 31st? Now find a formula for the total. Now find a formula for finding the circumference of the earth at your latitude. Now find the distance to the moon and back at perigee from your location. Now find… Never mind.

117. Belly button biodiversity

What is the average number of new-to-science microbes found per belly button?

2,368 species − 910 previously known = 1,458 new-to-science microbes

1458 ÷ 60 = **24.3 new-to-science microbes per belly button**

118. Heavy thinker

0.5 pounds compared to 3 pounds is 0.5/3.0 = 0.166666

Your brain size will have increased **approximately 17%.**

119. In the Aftermath Restaurant

How much did David's mom pay for the three of them?

Asa's order:		
Macaroni and Sneeze	$15.20	
Large Classic Choke with cup	$2.50	
Rhubarb wire pie—4 feet	$12.00	
Mary's order:		
Deep fried lint	$17.25	
with French flies	$4.50	
Sweep potato pie	$10.55	
Water	$0.00	
David's order:		
Thinsulate Pancakes	$14.68	
Large Arti Choke with cup	$2.50	
Last Stand Custard	$7.25	
	subtotal	$86.43
$86.43 * 0.05 = $4.32 sales tax		
$86.43 + 4.32 = $90.75 total bill		

120. Care to join us?

6 disgusting entrees * 6 repulsive beverages * 6 revolting

desserts = 6 * 6 * 6 = **216 luscious possibilities!**

Selected sources

Apollo 24/7. (2022). "10 Interesting Facts About Poop You Didn't Know." General Health. www.apollo247.com/blog/article/interesting-facts-about-poop

Artis Micropia. (n.d.). "Microworld: The Most Powerful Life on Earth." www.micropia.nl/en/discover/stories/microworld

Birdwatching HQ. (2024). "25 Most Common Spiders in Wisconsin." https://birdwatchinghq.com/common-spiders-in-wisconsin

Bryson, B. (2019). *The Body: A Guide for Occupants.* New York: Doubleday.

Castro, J. (2022). "11 Surprising Facts About the Circulatory System." Live Science. www.livescience.com/39925-circulatory-system-facts-surprising.html

Cleveland Clinic. (2024). "DNA, Genes and Chromosomes." https://my.clevelandclinic.org/health/body/23064-dna-genes--chromosomes

Environmental Protection Agency. (2020). "Agriculture and Aquaculture: Food for Thought." www.epa.gov/snep/agriculture-and-aquaculture-food-thought

Kaufman, D. and Thomas, W. (1990). *Dolphin Conferences, Elephant Midwives, and Other Astonishing Facts about Animals.* New York: TarcherPerigree.

Pester, P. (2024) "The Longest-Living Animals on Earth." Live Science. www.livescience.com/longest-living-animals.html

Maci. (2024). "300 Weird Facts." Facts.net. https://facts.net/weird-facts

Nash, B. (1991). *Mother Nature's Greatest Hits: The Top 40 Wonders of the Animal World.* Los Angeles: Living Planet Press.

National Human Genome Research Institute (2020). "Chromosomes Fact Sheet." www.genome.gov/about-genomics/fact-sheets/Chromosomes-Fact-Sheet

National Institute of Health (NIH). (2016). "Indigestion (Dyspepsia)." www.niddk.nih.gov/health-information/digestive-diseases/indigestion-dyspepsia/symptoms-causes

NOVA/PBS. (2008) "Leeches." Science Now. www.pbs.org/wgbh/nova/sciencenow/0305/01.html

Our World in Data. (n.d.). "Research and data to make progress against the world's largest problems." https://ourworldindata.org

Ray, M. (2022). "Make Peace with Your Microbiome." https://drmitraray.com/make-peace-with-your-microbiome

Sloat, S. (2017). "How Much Do Humans Eat?" Yahoo! News. www.yahoo.com/news/much-humans-eat-numerical-breakdown-181300224.html

Smithsonian Institute. (n.d.). "Bug Info." Encyclopedia Smithsonian. www.si.edu/spotlight/buginfo

US Forest Service. (2015). "Project Edubat." BatsLive. https://batslive.fsnaturelive.org/edubat

Washington State University. (2022). "Ask Dr. Universe." https://askdruniverse.wsu.edu

Special thanks

We owe anywhere from begrudging thanks to undying gratitude
to the following people. We'll let them figure out which.

Ben
Scruffy
Taco Bell
our sisters
Margaret!
Mom, Dad
Pooky the cat
Hormel Foods
Lyssa, proofreader
Blue Bunny Popsicles
Math Lovers Anonymous
Shotgun Eddy's Whitewater Rafting

Printed in the United States
by Baker & Taylor Publisher Services